Old King Cole
and other rhymes

Illustrated by Pamela Storey

Contents

GONDOLA

Old King Cole
Was a merry old soul,
And a merry old soul was he;
He called for his pipe,
And he called for his bowl,
And he called for his fiddlers three.

Every fiddler, he had a fiddle,
And a very fine fiddle had he;
Twee tweedle dee, tweedle dee,
  went the fiddlers.
Oh, there's none so rare
As can compare
With King Cole and his fiddlers three.

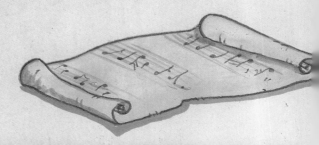

Little Boy Blue,
Come blow your horn,
The sheep's in the meadow,
The cow's in the corn;
But where is the boy
Who looks after the sheep?
He's under a haystack,
Fast asleep.
Will you wake him?
No, not I,
For if I do,
He's sure to cry.

Pussy cat, pussy cat,
   where have you been?
I've been to London
   to look at the queen.
Pussy cat, pussy cat,
   what did you there?
I frightened a little mouse
   under her chair.

Pat-a-cake, pat-a-cake, baker's man,
Bake me a cake as fast as you can;
Pat it and prick it, and mark it with B,
Put it in the oven for baby and me.

There was an old woman
who lived in a shoe,
She had so many children
she didn't know what to do;
She gave them some broth
without any bread;
She whipped them all soundly
and put them to bed.

Chook, chook, chook,
Good morning Mrs Hen
How many chickens have you?
Madam, I have ten.

Four of them are yellow,
Four of them are brown,
Two of them are speckled,
The nicest in the town.

This little pig went to market,
This little pig stayed at home,
This little pig had roast beef,
This little pig had none,
And this little pig cried,
Wee-wee-wee-wee-wee,
All the way home.

Mary had a little lamb,
Its fleece was white as snow;
And everywhere that Mary went
The lamb was sure to go.

It followed her to school one day,
That was against the rule;
It made the children laugh and play
To see a lamb at school.

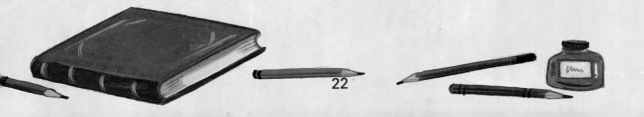

The north wind doth blow,
And we shall have snow,
And what will poor robin do then?
Poor thing.
He'll sit in a barn,
And keep himself warm,
And hide his head under his wing.
Poor thing.

Cock a doodle doo!
My dame has lost her shoe,
My master's lost his fiddlestick,
And knows not what to do.

I see the moon,
And the moon sees me;
God bless the moon,
And God bless me.